TEETH AND CLAWS

ANIMAL WEAPONS

Lynn M. Stone

The Rourke Press, Inc.
Vero Beach, Florida 32964

© 1996 The Rourke Press, Inc.

All rights reserved. No part of this book may be reproduced or utilized in any form or by any means, electronic or mechanical including photocopying, recording or by any information storage and retrieval system without permission in writing from the publisher.

PHOTO CREDITS
© Frank S. Balthis: p. 13; all other photos © Lynn M. Stone

Library of Congress Cataloging-in-Publication Data

Stone, Lynn M.
   Teeth and claws / Lynn M. Stone.
     p. cm. — (Animal Weapons)
   Includes index
   Summary: Describes the appearance and uses of different kinds of animal teeth and claws, with an emphasis on their utility as weapons.
   ISBN 1-57103-165-0
   1. Teeth—Juvenile literature. 2. Claws—Juvenile literature.
[1. Teeth. 2. Claws. 3. Animal weapons.]
I. Title II. Series. Stone, Lynn M. Animal weapons.
QL858.S765 1996
591.53—dc20   96–8996
                                                         CIP
                                                          AC

Printed in the USA

TABLE OF CONTENTS

Teeth and Claws	5
Tusks	6
Reptile Teeth	9
'Gators and Crocs	11
Furry Meat Eaters	14
Dog Family	16
Cat Family	19
Bear Family	20
Seal Family	22
Glossary	23
Index	24

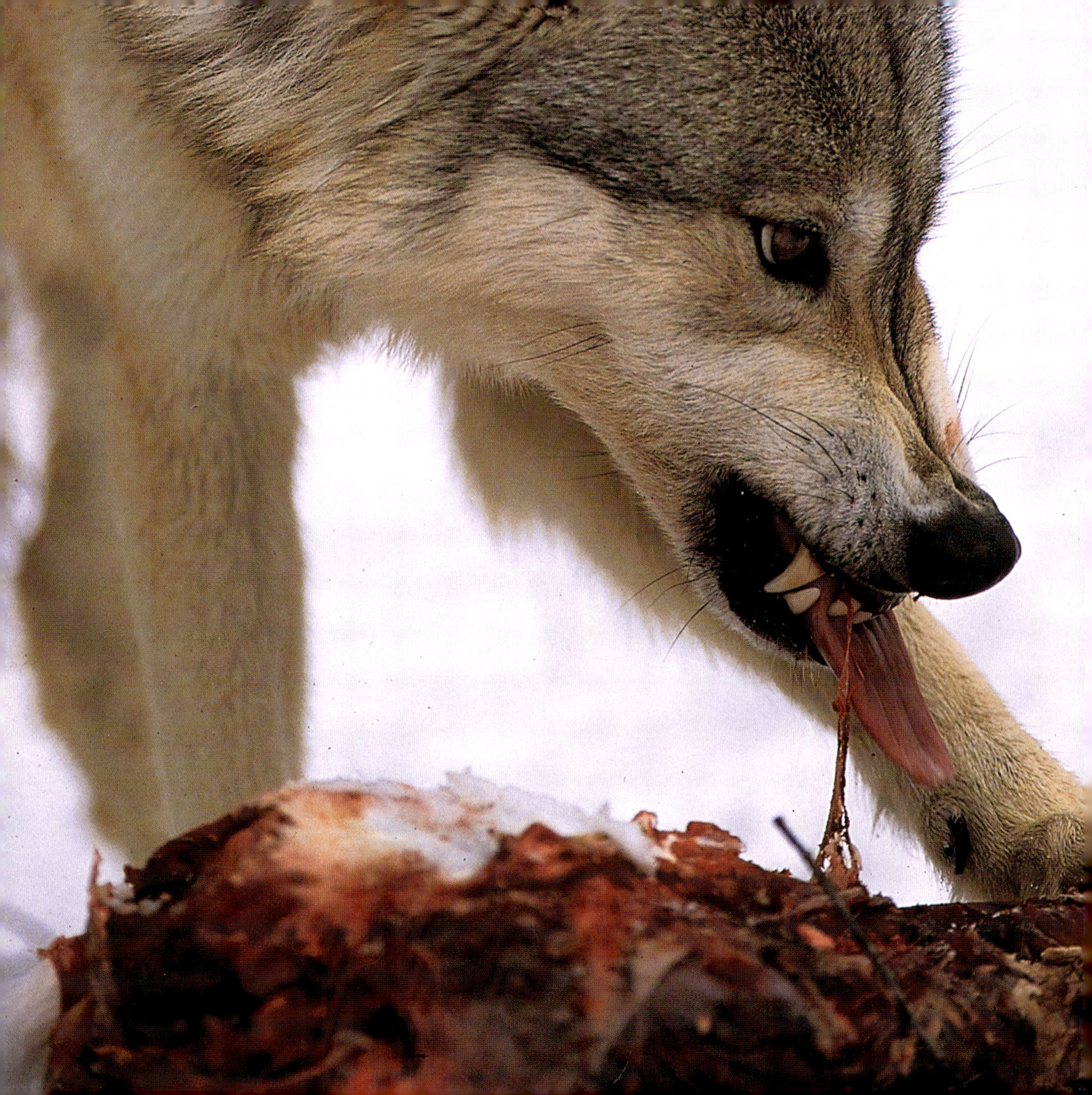

TEETH AND CLAWS

When a tiger or wolf snarls, it's showing anger. It's also showing its most deadly weapons—a **lethal** (LEE thul) set of teeth.

Many animals, such as tigers and wolves, depend upon teeth as weapons for killing. Their teeth and the teeth of other **predators** (PRED uh turz) are made for the work of hunters.

Many predators have sharp claws, too. The claws are weapons, but they are not as deadly as sharp teeth.

The teeth that killed the wolf's prey now threaten another wolf

TUSKS

Several animals have one or more extra-long teeth called tusks. Tusks are so long that they stick out of the animal's mouth even when it's closed!

Tusks are sharp, and they're often used as weapons. They are used for other purposes, too.

The longest tusks are the elephant's. They can be more than 10 feet long.

Walrus tusks point down. They can be more than three feet long. Wild pigs grow sharp tusks that sweep upward.

Warthogs of Africa defend themselves from big cats with sharp, slashing tusks

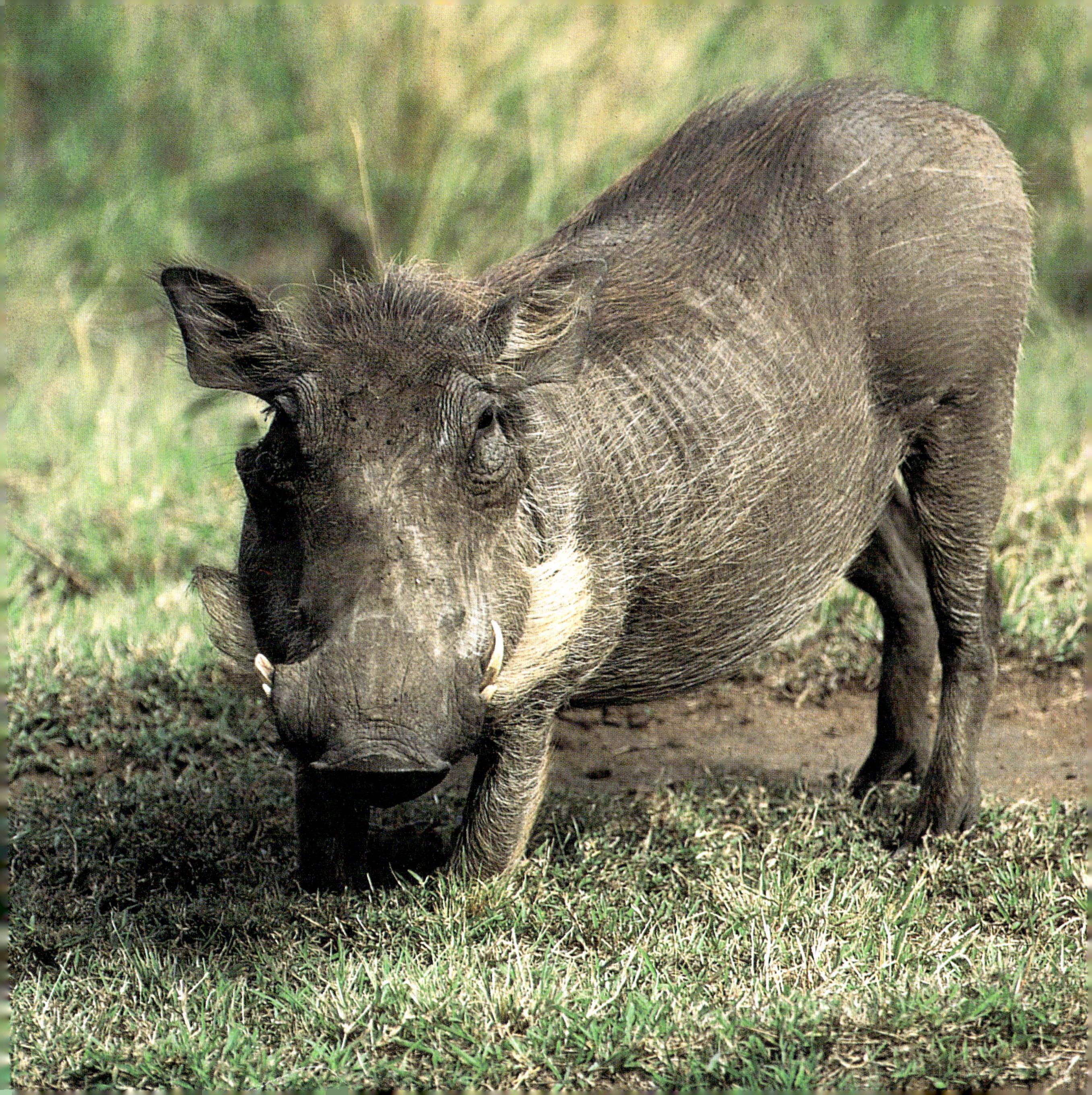

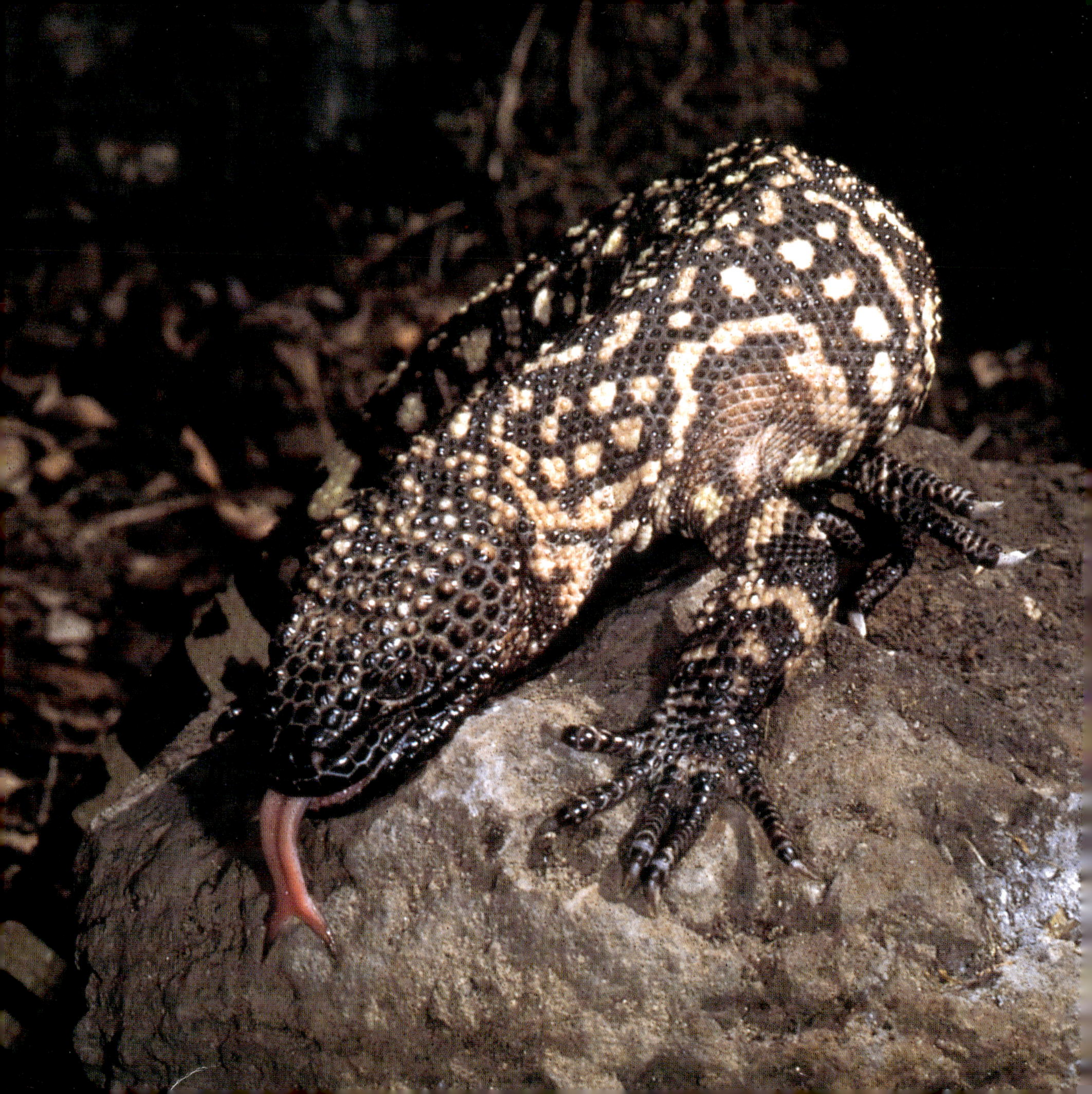

REPTILE TEETH

Crocodiles, alligators, snakes, and most lizards are meat-eating reptiles. They depend upon sharp teeth as weapons to grab or kill prey.

The teeth of snakes are slender and sharp. Several **venomous** (VEN uh mus), or poisonous, snakes have a pair of long, hollow teeth called fangs. The fangs hold **venom** (VEN um). Venom is a liquid poison.

Lizards have mouths full of sharp teeth. Only two kinds of lizards are venomous. They are the gila monster of the Southwest and the beaded lizard of Mexico.

The Mexican beaded lizard is one of just two venomous lizards in the world

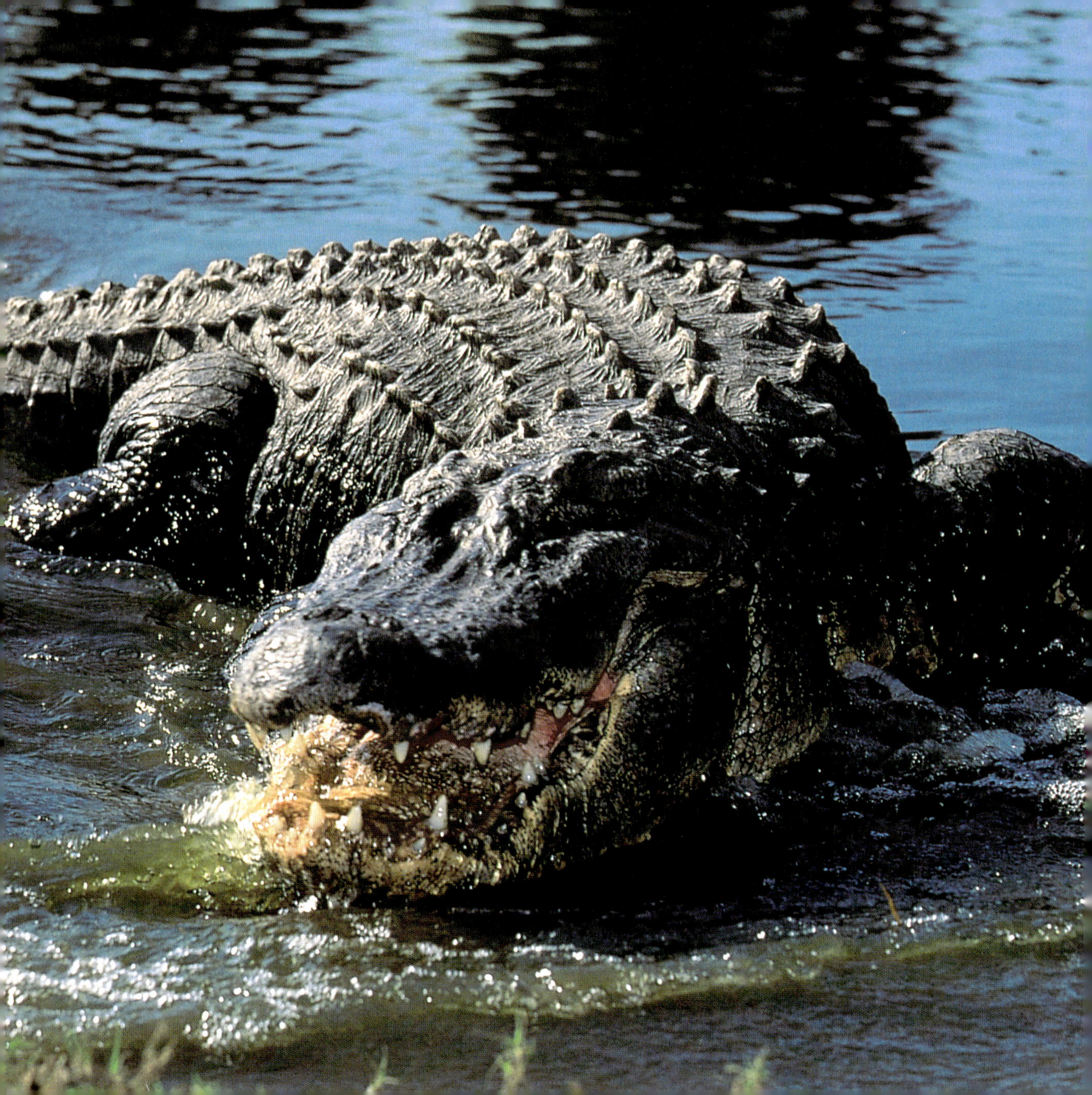

'GATORS AND CROCS

The **crocodilians** (KRAHK uh DIL ee unz) are a reptile family that includes crocodiles, alligators, and **gavials** (GAY vee ulz). They are the world's largest reptiles. A few reach 20 feet in length.

Crocodilians have long jaws with sharp, mostly cone-shaped teeth. Their teeth are designed for grabbing and holding prey rather then chewing or sawing it.

Crocodilian jaws have great power when they close. They easily crush turtle shells and snap legs of large animals.

Crocodilians grow new teeth to replace lost or broken ones.

The alligators toothy "grin" can be lethal when a 'gator lunges for prey

Ready to use teeth and claws, a cougar dashes across a Montana mountain meadow

A biting battle between northern elephant seal bulls turned into a blood bath. Despite bloodshed, bulls usually recover from their battles over female seals

FURRY MEAT EATERS

Many of the world's furry animals, the mammals, are meat eaters. Meat-eating teeth come in a variety of sizes and shapes. However, all meat eaters have some teeth that grab, tear, and cut.

The meat-eating mammal families with teeth include cats, dogs, bears, raccoons, hyenas, mongooses, otters, seals, certain whales, bats, and some pouched animals.

Most bears and some of the other meat eaters mix plant foods with their diets.

The teeth of a Canada lynx seize its favorite wintertime prey, a snowshoe hare

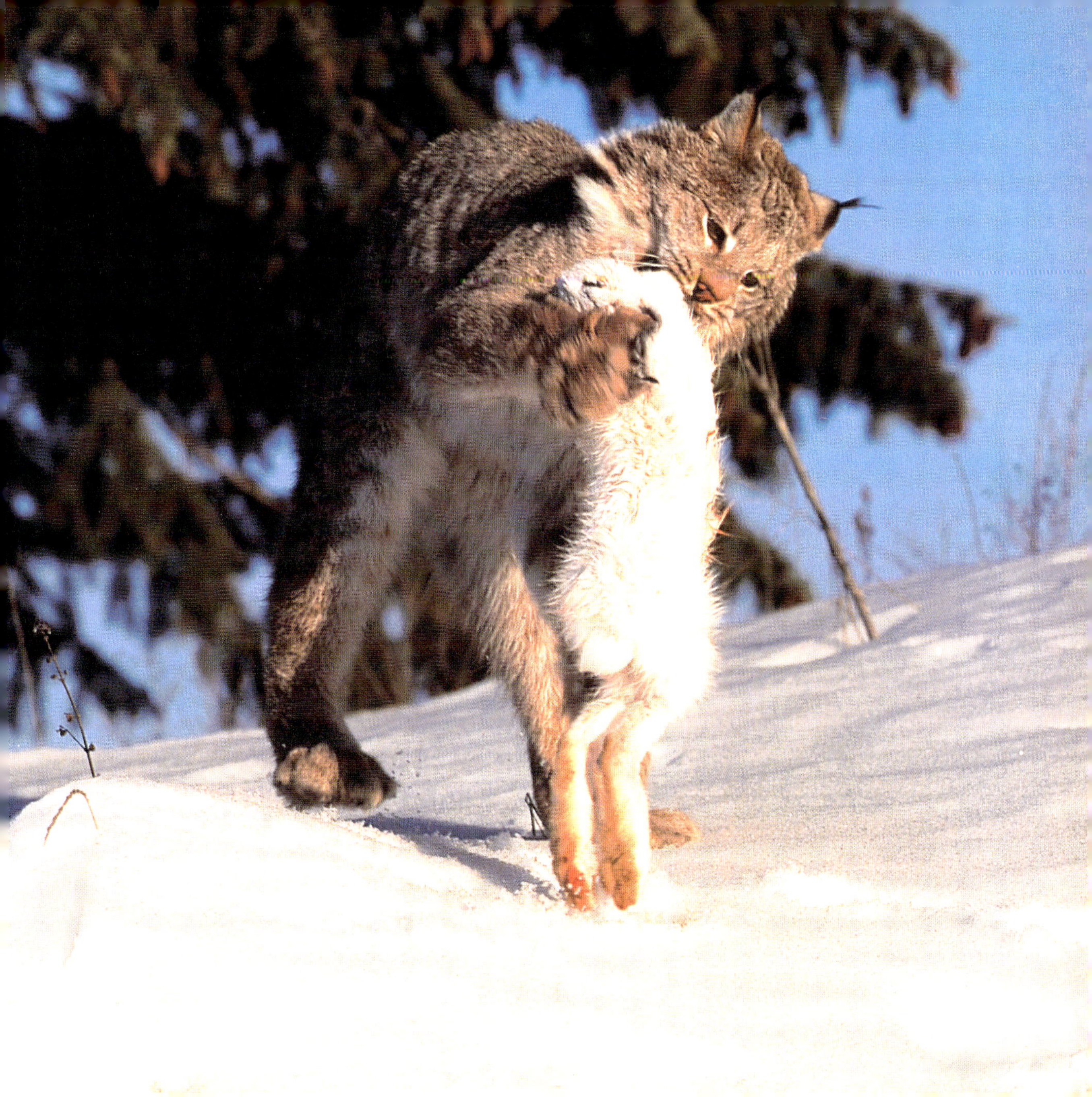

DOG FAMILY

The family dog's wild cousins are hunters. They kill to eat. They rarely eat plants or vegetables.

Foxes, coyotes, wolves, jackals, and African wild dogs kill prey by biting. Some of these wild members of the dog family hunt in packs. A pack can kill prey animals much larger than the hunters.

The claws of dog family members help them grip the ground or a struggling animal. Teeth are their most lethal weapons.

By hunting together as a pack, wolves can bring down caribou and other prey larger than themselves

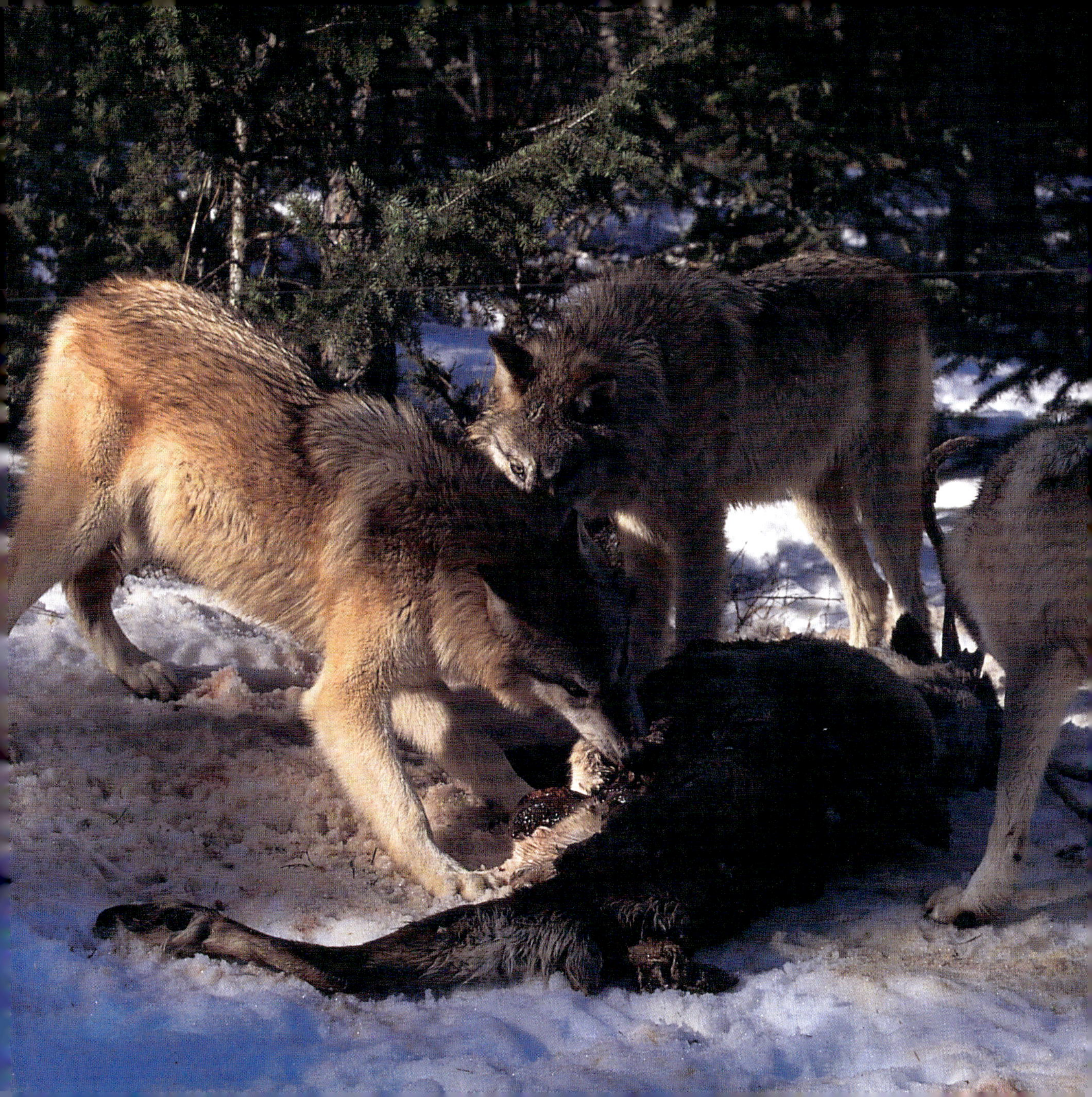

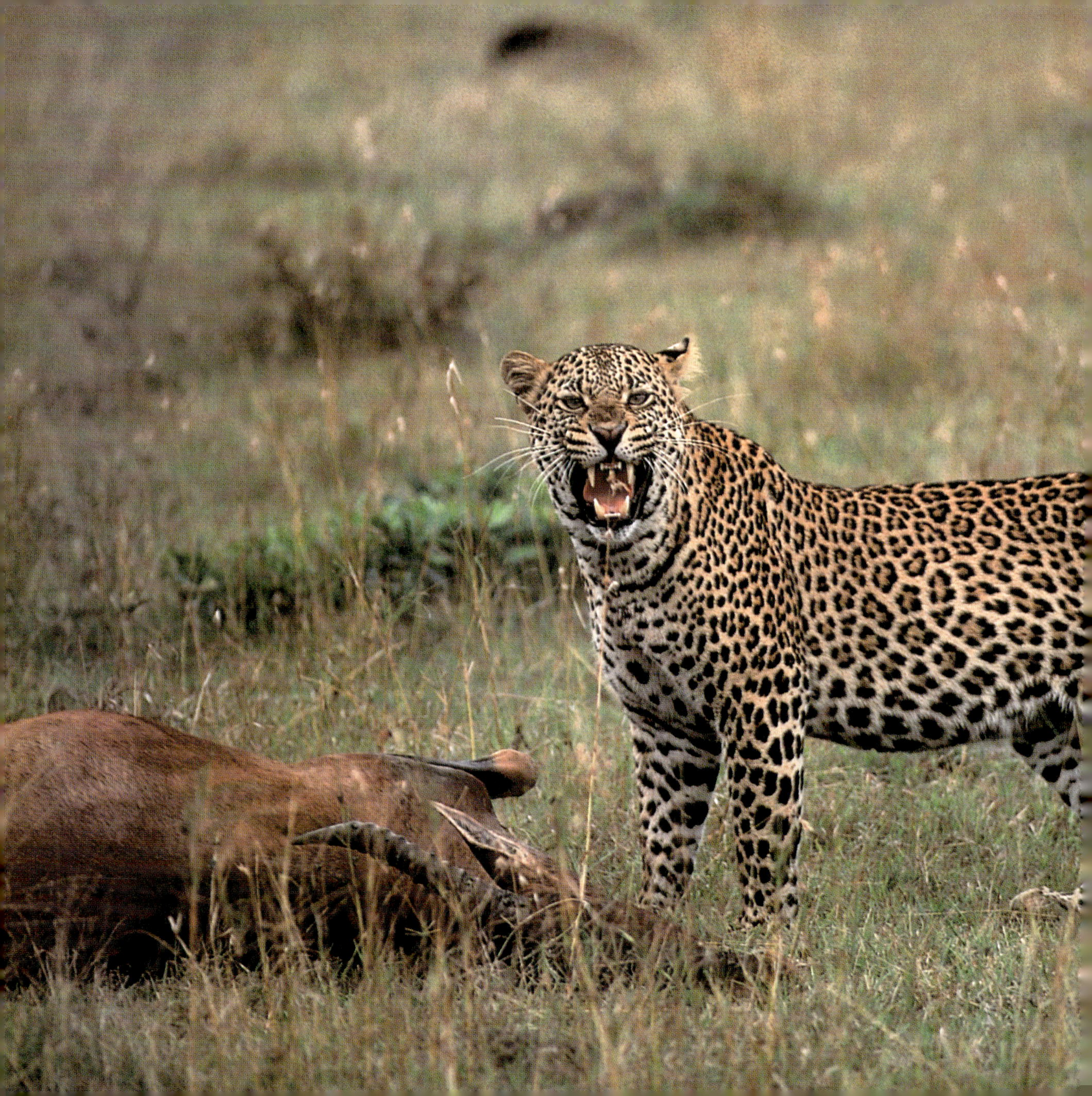

CAT FAMILY

Cats are perhaps nature's finest killing machines. They have keen senses, speed, and athletic bodies.

They also have fearsome teeth made to grab and cut. Their claws are sharp and deeply curved. Sharp claws help a cat haul down and hold prey.

Most kinds of cats keep their claws sharp by pulling them into their paws. A cat only extends its claws to attack or to sharpen them on a tree.

A leopard's snarl warns other predators away from its antelope kill in East Africa

BEAR FAMILY

The biggest meat eaters on Earth don't always eat meat. They are brown bears of Kodiak Island, Alaska. They can weigh nearly 2,000 pounds.

Kodiaks and other bears, except the polar bear, mix fruit, leaves, nuts, berries, and grass into their diets. Polar bears live almost entirely on flesh.

Bear teeth are useful for chewing plants as well as killing prey.

Bear claws are long and curved. They are made for digging and tearing.

Big Alaska brown bears sometimes turn their fangs and claws against each other in fights over fishing holes

SEAL FAMILY

Most seals catch and eat fish. They're equipped with big, sharp teeth for the job.

Seals of one kind or another live throughout the world's oceans. Most seals live in the cold seas of the Arctic and Antarctic.

Male, or bull, seals sometimes turn their teeth against each other. They fight bloody battles for the right to take female seals as mates.

Glossary

crocodilian (KRAHK uh DIL ee un) — the reptile family of crocodiles, alligators, and gavials; large, long-snouted reptiles

gavial (GAY vee ul) — a large reptile similar to the crocodile, but with a very narrow, toothy snout

lethal (LEE thul) — very dangerous; deadly

predators (PRED uh turz) — animals that hunt other animals for food

venomous (VEN uh mus) — able to produce venom; poisonous

venom (VEN um) — a poison produced by certain animals, including several snakes, fish, and spiders

INDEX

alligators   9, 11
battles   22
bears   14, 20
   brown   20
   Kodiak   20
   polar   20
cats   14, 19
claws   5, 16, 19, 20
crocodiles   11
crocodilians   11
dogs   14, 16
gavials   11
gila monster   9
lizards   9
   beaded   9
mammals   14

packs   16
predators   5
prey   9, 16, 19, 20
reptiles   9, 11
seals   14
snakes   9
   venomous   9
teeth   5, 6, 9, 11, 16, 19, 22
tusks   6
venom   9
weapons   5, 6, 16